dioptase,
Congo

pyrite,
Illinois, USA

calcite,
Cumbria, UK

stibnite,
Japan

These are natural crystals. They formed within the Earth's crust or upon its surface. Recovered from the depths of mines and caves, or from volcanoes, these crystals have not been cut or polished: this is the way they grew. Their crystal faces, shapes and colours differ because they are chemically distinct materials whose atoms are arranged in specific three-dimensional patterns. Solids with an ordered atomic structure are described as crystalline. Virtually all inorganic chemicals are crystalline, as are many organic materials, including proteins, vitamins and even some viruses. Nowadays, many different crystals can be synthesized for uses ranging from electronic components to medicines. A good way to start exploring crystals is to observe the wealth of varieties produced by nature. This book is designed to help you discover more about crystals, their beauty, properties and uses.

fluorite,
Switzerland

microcline,
Colorado, USA

hemimorphite,
Cumbria, UK

ENQUIRE WITHIN

There are many thousands of different orderly ways that atoms can be arranged in crystals, yet all of these fall into just 14 basic lattice patterns (fig 7). The repeating pattern of atoms in a crystal can be represented simply by an array of lattice points. For a particular crystal structure, each lattice point is surrounded by exactly the same geometrical arrangement of atoms. The structure is described by the repetition of this atom arrangement at each lattice point in three dimensions.

The orderly atomic framework inside a crystal also governs its external shape. As a crystal 'grows', atoms add on in ordered fashion, forming flat layers called faces. Faces become overgrown as the crystal increases in size, and successive growth surfaces may be visible inside transparent crystals, preserved as differently coloured layers (figs 5, 6). Crystalline extra-terrestrial materials impact the Earth as meteorites; the cut surface of an iron meteorite reveals an intergrowth of iron–nickel alloy crystals (fig 3).

- ● oxygen
- • hydrogen

1 Crystal structure of ice.

3 Metal crystals in an iron meteorite.

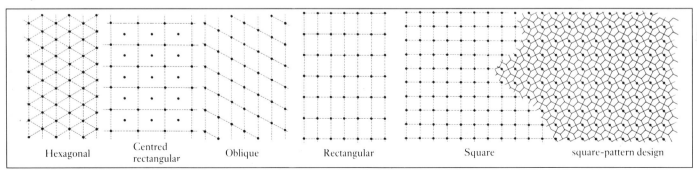

| Hexagonal | Centred rectangular | Oblique | Rectangular | Square | square-pattern design |

2 Five basic regular, flat lattice patterns are possible in two dimensions. The added design (right) has an identical motif geometry at each lattice point.

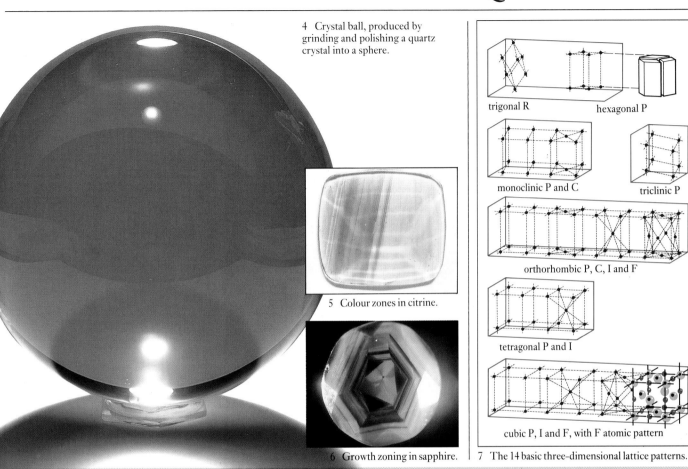

4 Crystal ball, produced by grinding and polishing a quartz crystal into a sphere.

5 Colour zones in citrine.

6 Growth zoning in sapphire.

trigonal R

hexagonal P

monoclinic P and C

triclinic P

orthorhombic P, C, I and F

tetragonal P and I

cubic P, I and F, with F atomic pattern

7 The 14 basic three-dimensional lattice patterns.

Atoms, structures and lattices

Although there are countless regularly repeating two-dimensional designs (often used for wallpaper), all conform to just five basic types of lattice pattern (fig 2). Crystals are natural 'designs' of atoms in regularly repeating three-dimensional arrays called crystal structures. The precise crystal structure adopted by atoms depends upon the temperature, pressure and the kinds of atoms incorporated into the growing crystal.

For example, a crystal of sapphire weighing just one gram contains 30,000 million million million atoms of aluminium and oxygen that have bonded at high temperature. Iron and titanium atoms, incorporated during growth, cause different depths of blue colour in sapphire (fig 6).

All the different crystal structures known can be assigned to one or other of only 14 lattice patterns (fig 7). These basic patterns relate to the way atoms bind together, and thus control the crystal shapes (pages 6, 16). Joining lattice points by straight lines generates unit cells. Lattices are classified according to their unit cell shape and whether lattice points occur only at cell corners (P = Primitive, R = Rhombohedral), or, additionally, at the centres of opposite sides (C = Centred), at the body centre of the cell (I = German *Innenzentrierte*), or at each face of the cell (F = Face Centred).

ON THE OUTSIDE

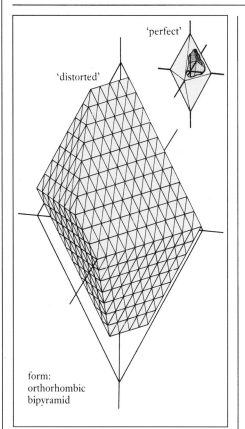

'perfect'

'distorted'

form:
orthorhombic
bipyramid

8 Crystals grow in planes parallel to each face. Angles between faces are not altered by distorted growth.

Crystal faces grow in a puzzling variety of shapes and sizes. Even the same crystalline substance can form crystals of very different shape (fig 9). To add to the confusion, many crystals grow with distorted-looking shapes – but the angles between faces are set by the three-dimensional pattern of the atomic structure within. Crystals with distorted outer shapes can have the same degree of atomic order inside as a perfect-looking crystal.

A crystal face is a final growth surface. Crystals grow when atoms add onto their outer surfaces (earlier faces) in parallel layers. Each layer added repeats the regular pattern of atoms in the structure.

A crystal is produced by growth of a set of planes in different directions, parallel to each face (fig 8). As several sets of planes build outward (fig 11), the developing crystal shape depends upon the conditions during growth. The relative growth rates of each face determine their relative sizes and can create alternative finished shapes or habits (figs 10, 15).

12 Cleavage planes in baryte: cracks in two directions.

9 Quartz: different shapes; same angles.

10 Crystal habits of pyrite.

11 Slower build-up leads to larger faces.

4

13 Crystal shapes of some minerals.

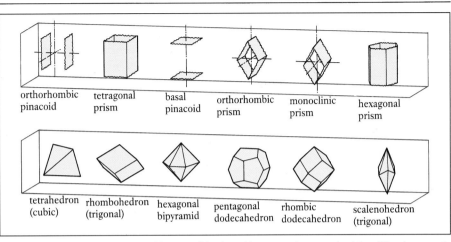

14 Examples of forms: related sets of faces are either 'open' (cannot enclose space) or 'closed' (enclose space).

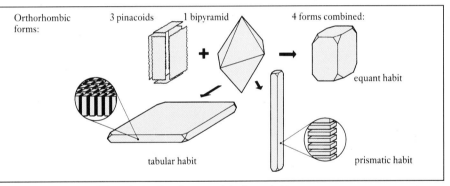

15 The same forms, if differently developed, can result in diverse habits.

Crystal shapes

Crystals grow with plane faces, and these tend to develop parallel to surfaces containing strongly bonded atoms (fig 13). Planes with the same atomic pattern develop as a number of equivalent faces, and together these are called a form (fig 14). Crystal shapes can be described as bounded by one or more forms. The main faces that develop are usually those parallel to planes containing the highest density of lattice points (fig 7), so the unit cell shape is often related to the form developed. For example, crystal structures with long thin unit cells often form tabular crystals (fig 15). However, the growth environment also controls variations in form, habit and distortion, so interpreting crystal shapes is far from easy!

Although face sizes and shapes vary in real crystals, the angles between corresponding faces are identical in *all* crystals of the same substance and structure, regardless of how they grew. The angles are set by the internal structure. The atomic structure also controls the orientation of cleavage: some crystals split more easily parallel to certain planes, and these correspond to layers of weaker bonding between the atoms (fig 12). The angles between cleavage planes in different directions are governed by the internal atomic structure.

DIVERSE IMPERFECTIONS

Real crystals rarely have 'perfect' shapes like the drawings and models shown in books. Perfect crystal shapes are a useful aid in helping to understand the diversity of imperfections that abound in real crystals. The 'imperfect' habits of many natural crystals (fig 16) have a special fascination and may be extremely beautiful. Examples are to be found in the mineral collections of many famous museums.

Crystals grow in whatever way they can, and may trap impurities or other tiny crystals within them as they grow. Fluids can also become trapped within a crystal; millions of microscopic cavities containing trapped fluid can give quartz a 'milky' appearance (fig 18). Some crystals grow more easily when they share a plane of atoms with another crystal of the same structure growing in a symmetrical direction; these are called twinned crystals (figs 17, 21, 23, 24). Other crystals may form radiating clusters by the simultaneous growth of thousands of tiny crystals in unrelated directions (fig 19). Such needle-shaped crystals result if, for some reason, atoms become attached more easily to the crystal structure in one particular direction.

18   Trapped bubbles make quartz look 'milky'.

19   Mesolite: multiple growth of crystal needles.

16   A gypsum crystal that grew under stress.

17   Crystal layers in twinned feldspar.

20   Diamond with curved-looking faces.

21 Twinned cerussite crystals produce a star-shaped group.

Common crystal imperfections
Attempts by different planes to form their own faces often lead to tiny steps between alternating planes forming striated or apparently curved faces (fig 20). Regions of a crystal face growing at different rates may produce hollow or 'hopper' faces (fig 25), or tree-like branching 'dendrites' (fig 22). Twinned crystals can result from the geometry of the initial atom cluster, or during growth as conditions change.

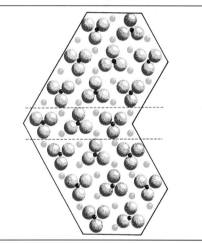

24 A twinned crystal structure.

22 'Dendritic' growth of pyrolusite.

23 'Skeletal' growth of twinned cerussite.

25 Salt crystals with stepped 'hopper' faces.

CRYSTAL WORLD

Planet Earth is made up mostly of crystals. We can only study directly the crystalline materials that occur near to the Earth's surface – a very small fraction of its mass. Most rocks of the Earth's crust consist of silicate crystals, in which oxygen and silicon atoms predominate. Chemically, the Earth's crust is estimated to be about 75% by weight oxygen and silicon, with only six other elements bringing the total to 99%.

Crystals of water (ice and snow) are constantly melting and reforming at the Earth's surface and in the Earth's atmosphere (fig 26). Crystals also grow in biological systems and are essential to many life forms.

26 Ice crystals: the 60-degree angles are governed by the internal pattern of oxygen and hydrogen atoms.

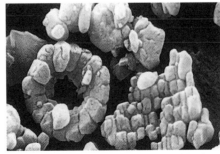

27 Crystals of calcium carbonate secreted by algae. Chalk rock is made up of billions of these tiny crystals.

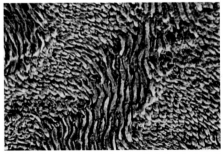

28 Tooth enamel: a tough mass of calcium hydroxyl phosphate crystals, which also form a part of bone.

29 Rock fragments detached by erosive forces, abraded and water-worn into pebbles on a beach. The different minerals in these pebbles show up as different colours.

Crystal chemistry on Earth

Oxygen atoms in rocks account for nearly half the total weight of the Earth's crust. The second most abundant element in the crust is silicon, which bonds very effectively with oxygen to form silicates. Most rock-forming minerals are silicate crystals, and their huge variety is made up of combinations of just a few main elements: oxygen, silicon, aluminium, iron, calcium, sodium, potassium and magnesium. Life forms also make use of these elements, together with essential carbon, hydrogen and phosphorus. Biologically formed crystals of calcium carbonate (fig 27) have accumulated in their billions to form limestone rocks such as the white chalk cliffs of Dover, UK, while calcium hydroxyl phosphate (apatite) crystals form important components of teeth and bones (fig 28). Oxygen atoms are larger than most other atoms, so they occupy the bulk of the volume in oxide and silicate crystal structures. Many common crystals consist of closely packed oxygen atoms with the small spaces between them occupied by smaller atoms such as silicon. The crystal structures of rock-forming minerals are largely determined by the internal pattern adopted by oxygen atoms. Oxygen and hydrogen form an important chemical on Earth – water, which crystallizes at 0°C.

MINERALS

A *mineral* is a natural crystalline substance with a definite chemical composition. Crystals with structures identical to minerals can be reproduced in a chemical laboratory, but these are not minerals. Crystals produced by life forms, although natural, are not normally classified as minerals.

Minerals make up the rocks of the Earth's crust and its deep inner layers. The most common rock-forming minerals are made up of linked units of strongly bonded silicon and oxygen atoms (fig 33). Quartz contains just silicon and oxygen; other more complex silicates may contain a variety of additional elements such as aluminium, calcium, sodium, magnesium or iron. Metals are extracted from ore minerals, usually metal sulphides or oxides, sought for their economic concentrations of certain metals.

Ice and diamond are two important non-silicate minerals, formed in very different environments. An important carbonate mineral, calcite, forms much of the world's limestone and marble, and is a temporary store for atmospheric carbon dioxide, in the form of calcium carbonate, at the Earth's surface.

31 Common rock-forming minerals.

30 Granite (left): an intergrowth of quartz, feldspar and mica. Sandstone: detrital quartz and rock fragments. Schist: minerals recrystallized under pressure.

Rock-forming minerals

Over 3600 different minerals have so far been discovered, but most of these are very rare. Only 30 or so are common at the Earth's surface. Minerals that make up the Earth's rocks are called rock-forming minerals (figs 30, 31); most are silicates. Some rocks, such as pure quartz sandstone or marble, are made up from crystals of just one mineral. However, most rocks are aggregates of two or more minerals. Igneous rocks form when minerals crystallize from molten rock (magma) as it cools from high temperatures, and the cooling rate determines the size of the crystals (fig 32). Metamorphic rocks have been subjected to increased pressure and temperature conditions, causing their mineral structures to be rearranged atom by atom in the solid state.

The basic atomic unit in the architecture of all silicate structures is a tetrahedron of four oxygen atoms with a silicon atom at its centre (fig 33). These tetrahedral units link together in many different ways to form a wide range of silicate minerals.

32 Crystals of iron–magnesium silicate are enclosed by small feldspar crystals (inset) as this lava cools.

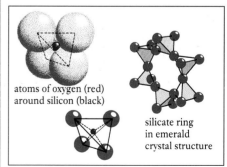

atoms of oxygen (red) around silicon (black)

silicate ring in emerald crystal structure

33 Silicon–oxygen units and a silicate structure.

GEM QUALITY CRYSTALS

Certain minerals and some man-made crystals are cut and polished into gemstones. To qualify as such, they should be hard and tough, be 'crystal' clear and have a beautiful colour. Rarity or special physical properties will enhance the value of gemstones, as will fashion. The majority of gemstones are silicates (e.g. emerald and amethyst). Ruby and sapphire are both composed of aluminium oxide (corundum) with different kinds of impurity atoms. Diamond is composed of a single element – carbon, and is the hardest substance known. A skilled lapidary can turn a rough-shaped crystal into a sparkling gem by carefully cutting and polishing facets around the stone at angles designed to reflect the maximum amount of light entering the gem (figs 34, 38). The interaction of light with a gem (its optical properties, see page 26) provides a means of identifying the crystalline substance from which the gem was cut (figs 35, 36).

35   Double refraction of light by sinhalite gem.

36   Varied colour effect as an iolite gem is turned.

34   A faceted tanzanite cut from a gem quality crystal.

37   Jade: tough rock of interlocked crystals.

38   Gem quality crystals, faceted to reveal their clarity, colour and beauty.

Crystal chemistry of gems

Many silicate and oxide minerals are hard, transparent and resistant to wear and corrosion. These qualities make them suitable for use as gemstones. However, the colour of the mineral often determines whether it will be used as a gemstone. For example, the mineral beryl – a beryllium aluminium silicate – is colourless, but a tiny amount of chromium in its structure (in place of some of the aluminium atoms) can make it green – this is emerald. But chromium impurity atoms in the structure of corundum (aluminium oxide) produce the red colour of ruby. Differences in the positions of the oxygen atoms surrounding chromium in these two structures affect the way the white light interacts with the electrons of the chromium atom. The blue colour of sapphire is caused by the incorporation of both iron and titanium impurities in corundum. Different kinds of impurity atoms can produce a wide range of colour variations even in the same crystal structure. Natural gemstones are often heat-treated in order to induce colour changes to make them more striking in appearance. Many laboratory-grown crystals with the same chemistry and structure as minerals are cut as 'synthetic' gems. Other crystals can be made that have no natural counterpart: these are cut as 'imitation' gems.

SYMMETRY

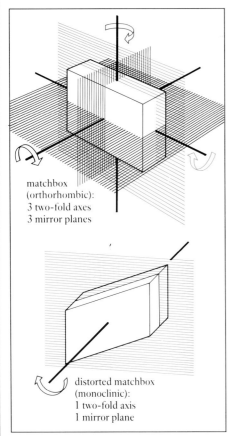

matchbox
(orthorhombic):
3 two-fold axes
3 mirror planes

distorted matchbox
(monoclinic):
1 two-fold axis
1 mirror plane

39 Two solid shapes and their symmetry.

Crystal shapes, atomic structures and lattice point arrays can all be described and classified according to the symmetry they possess. All regular patterns can be described in this way. Many familiar shapes have recognizable kinds of symmetry (fig 39), such as symmetry planes that, if they were mirrors, would reflect the object exactly, or axes around which the object takes up an identical orientation several times in a complete rotation. Crystals that exhibit high symmetry possess many planes and axes of symmetry (fig 41). Crystals of low symmetry possess only a few symmetry elements.

The symmetry of ideal crystal shapes can be readily determined. However, symmetry-related crystal faces will only be of equal size if the crystal grew under ideal conditions. Real crystals are rarely well-formed, and their external shape may not always be symmetrical. But measurement of the angles between faces can reveal the true symmetry of a crystal, regardless of its growth distortion. The angles between faces are governed by the atomic structure within (see pages 2 and 4), and the repeat pattern of the atomic motif itself can be described by the same symmetry rules.

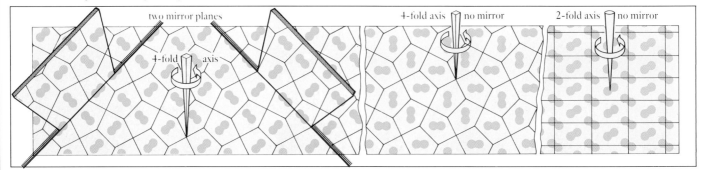

two mirror planes

4-fold axis

4-fold axis ☐ no mirror

2-fold axis ☐ no mirror

40 Differences in these patterns can be described precisely by the symmetry elements they possess.

one set of
three axial
planes

three sets of
two diagonal
planes

centre

three
four-fold axes

four
three-fold axes

six
two-fold axes

Atoms in a cubic crystal:
sodium (red) and chlorine (blue)
in halite (common salt).

41 The cubic crystal structure of sodium chloride, and its symmetry elements.

Crystallographic symmetry elements

Solid shapes or regular patterns can be described by the symmetry elements they possess (figs 40, 41). A symmetry operation moves (transforms) the pattern in such a way that it coincides with itself. For example, if a pattern is turned about a straight line (an axis), and it appears identical every 180, 120, 90 or 60 degrees, then the line is an axis of rotation.

Rotation axes are denoted as two, three, four or six-fold axes. If the pattern is exactly reflected (as in a mirror) on either side of an imaginary plane, then this is a mirror plane of symmetry. A centre of symmetry inverts each part of the pattern to an identical equidistant point in a straight line though the centre. Three-dimensional patterns, like atomic structures, may also have other symmetry elements whereby rotation or mirror

reflection is coupled with translation along the axis or plane by a set fraction of the pattern repeat. Rotation axes can also be coupled with inversion through a centre. All these crystallographic symmetry elements can be combined in only 230 different ways; these are known as the space groups. The tens of thousands of known crystal structures can each be assigned to one or other of these 230 space groups.

CRYSTAL SYSTEMS

Crystals are grouped into systems according to the symmetry of their external shapes. The internal lattice pattern of a crystal is related to its external shape (page 4). Angles between rows of lattice points are characteristic for each crystal system (fig 42). The presence or absence of certain basic kinds of symmetry element can be used to assign a crystal to one of seven crystal systems. The basic symmetry element required by a crystal for it to be placed in one or other of these systems is indicated in fig 43.

This basic symmetry is the minimum required for the system, but additional symmetry elements may be present in different crystals belonging to the same system. The maximum symmetry in each system is that of the lattice array (fig 7). For example, crystals of beryl and apatite both belong to the hexagonal system (they both have a six-fold rotation axis), but the atomic motif around each lattice point in beryl has a higher symmetry (additional mirror symmetry) compared with the atom array in apatite. So beryl and apatite grow crystals of different symmetry and belong to different classes of the hexagonal system. There are a total of 32 different crystal classes; all crystal shapes conform to one or other of these.

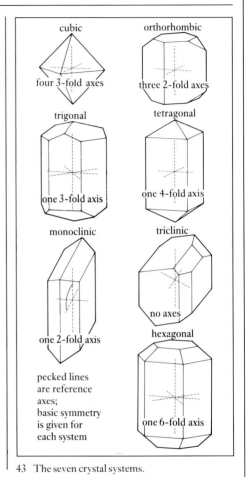

43   The seven crystal systems.

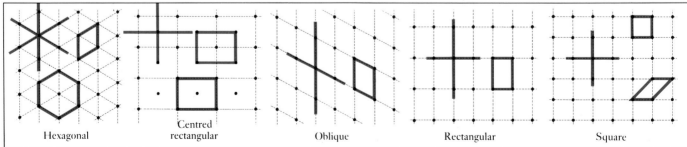

42   Planes of lattice points shown in two dimensions. Reference axis directions are drawn parallel to rows of lattice points.

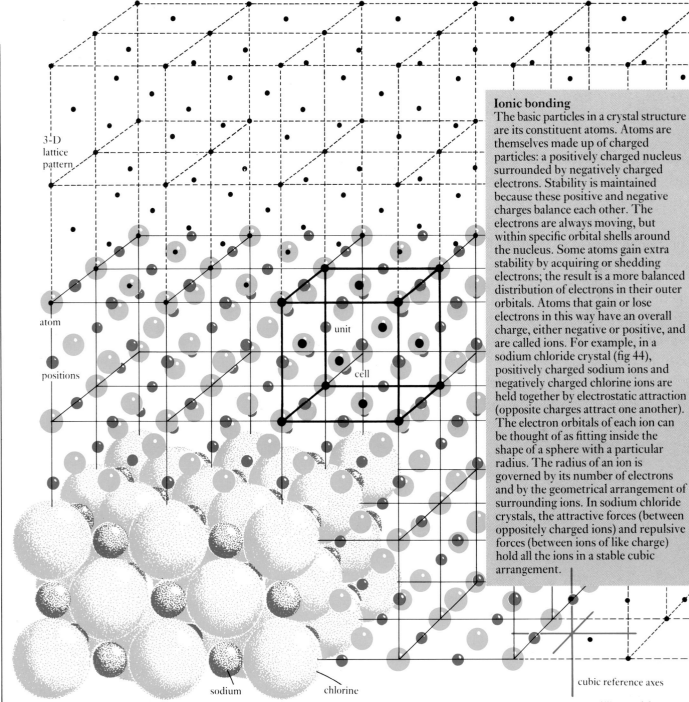

3-D lattice pattern

atom

positions

unit

cell

Ionic bonding

The basic particles in a crystal structure are its constituent atoms. Atoms are themselves made up of charged particles: a positively charged nucleus surrounded by negatively charged electrons. Stability is maintained because these positive and negative charges balance each other. The electrons are always moving, but within specific orbital shells around the nucleus. Some atoms gain extra stability by acquiring or shedding electrons; the result is a more balanced distribution of electrons in their outer orbitals. Atoms that gain or lose electrons in this way have an overall charge, either negative or positive, and are called ions. For example, in a sodium chloride crystal (fig 44), positively charged sodium ions and negatively charged chlorine ions are held together by electrostatic attraction (opposite charges attract one another). The electron orbitals of each ion can be thought of as fitting inside the shape of a sphere with a particular radius. The radius of an ion is governed by its number of electrons and by the geometrical arrangement of surrounding ions. In sodium chloride crystals, the attractive forces (between oppositely charged ions) and repulsive forces (between ions of like charge) hold all the ions in a stable cubic arrangement.

cubic reference axes

sodium

chlorine

44 The crystal structure and F-centred lattice (see fig 7) of sodium chloride. The lower part shows the relative sizes of the atoms as a space-filling model.

EVERYDAY CRYSTALS

We constantly encounter crystals, whether in the home, in the city, or at work. Crystals are part of everyday life and are used in a multitude of ways: they cool our drinks, make paper glossy, make concrete 'set', give strength to bricks, or enable computers to work. Ceramic pots and plates, metal pans and spoons, paints, cosmetics, salt, sugar, vitamin pills and painkillers – all these are crystalline. We rely on crystals in so many different ways, and much research goes into discovering how atoms are arranged in crystals so that we can understand why a crystalline material behaves as it does. Crystal structures can be designed to have special properties with new technological uses in mind.

Natural crystals and rocks are to be seen all around us, in building stones (fig 46), construction materials, carvings, jewellery and ornaments (fig 45). Some crystals are grown daily in the kitchen, such as ice in the freezer and 'fur' in the kettle (fig 47), and we add crystals to our food (e.g. salt, sugar, monosodium glutamate). Chocolate is also crystalline (page 40). Our bodies are also constantly re-modelling our bones by slowly dissolving and re-growing new bone apatite crystals.

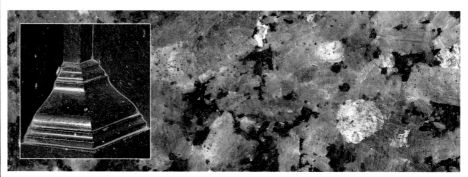

45 Agate snuff bottle, Chinese, 19th century.

46 Larvikite, a rock with reflective feldspar crystals, is commonly used to face buildings.

47   Calcium carbonate crystals growing from 'hard' water 'fur up' kettles and pipework (magnified 850 times).

CRYSTALS EVERYWHERE

Crystals grow everywhere, deep within the Earth and upon its surface. Some crystals may take millions of years to grow, and may be altered many times before you eventually see them at the Earth's surface. Mountains are made of crystals, billions upon billions of them interlocking together to form rocks. Crystalline rocks form at great depths; some of these are slowly brought to the surface by colossal forces within the Earth and are exposed as overlying rocks are steadily removed by water and wind erosion at the Earth's surface (fig 48).

48 Crystalline rock masses from deep within the Earth may eventually rise as overlying rocks are removed by erosion at the surface.

CRYSTALS EVERYWHERE

As rock fragments are detached from mountains and cliffs by ice, wind and water, some may be changed into a mixture of other minerals by reaction with water – most clay minerals form this way. More resistant rocks are rounded into pebbles (fig 51). The pebbles, mud, sand and clay are transported by water, wind or ice to new locations where they settle to form the beginnings of new rock formations. Some crystals from eroded rocks dissolve in water and are carried as ions in solution; eventually these may end up in the sea – keeping it salty. Saline water trapped in lakes may dry up and salts then crystallize from solution on the lake bed (fig 49).

Meteorites that impact on the Earth from space are samples of crystals from other bodies in the solar system (fig 50). The sizes of meteorites vary from dust particles to large rock fragments that can cause massive impact craters.

49 Lake salt.

50 This meteorite fell at Barwell, UK.

51 Pebbles: water-worn rocks.

BIRTH AND SURVIVAL

Crystals grow with particular atomic structures in response to the conditions of their growth environment. For example, diamond crystals grow naturally or in the laboratory from a supply of carbon atoms under great pressure and high temperatures. Diamonds form at depths greater than about 150 km in the Earth (fig 52). At 150–250 km depth the Earth's mantle is sometimes melted at localized 'hot spots' to form magma. The diamonds do not melt, but are incorporated into the magma, which, being buoyant, rises, taking the diamonds with it into the Earth's crust. During its ascent the magma also transports with it chunks of mantle rock – these often contain large crystals of garnet that grew at high pressures and temperatures. A few kilometres below the Earth's surface the magma solidifies to form kimberlite rock. Sometimes it erupts onto the surface, forming volcanoes (fig 53). After millions of years of erosion, diamond-bearing rocks are exposed. If eroded by water, then diamonds may accumulate in river gravels (fig 54).

When a crystal is removed from its birthplace, new conditions might destroy or alter it in some way. Diamond survives well because it is so hard, but if heated at low pressure it will turn into graphite as its atoms rearrange. Magma that rises to new levels in the Earth's crust may mix with other magma or encounter new conditions. Crystals that have already started to grow from the magma may then start to dissolve; these may be preserved as large corroded crystals trapped in a matrix of small crystals if the magma cooled quickly in a lava flow (fig 55).

52 Diamond from 250 km underground.

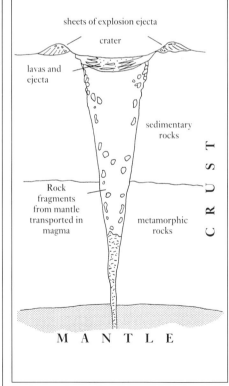

53 Kimberlite forms cylindrical pipe-like bodies.

54 Diamonds in river gravel.

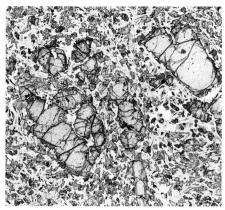

55 Corroded crystals in lava.

56 Crystals (inset) contained in volcanic ash (main picture).

Hot inside

Our planet Earth has a radius of 6370 km, and consists of three different layers – the crust, the mantle and the core. The outermost layer is the crust, which is up to 70 km thick in some places. The vast layer below this and extending down to 2900 km is the mantle.

The Earth is hot inside. At depths of 200 km the temperature is over 1400°C, and the pressure is enormous. This is where diamonds originate. The crystalline rocks making up the Earth's mantle are constantly re-crystallizing and moving in response to high temperature convection currents. These cause the upper parts of the mantle and crust to circulate slowly. Different regions (known as plates) circulate by moving in different directions, crushing against one another. Some plates are dragged under other plates as they collide, and other plates are crumpled as they ride over a descending plate. This constant slow movement is known as plate tectonics. In certain places the heat at depth within the Earth causes rocks to melt, and the magma produced may rise and cool to form new crystalline rocks below the surface. Some magma reaches the surface, creating volcanoes, scattering dust and crystals (fig 56) in explosive events, or pouring out as molten lava.

MICROSCOPY

Many crystals are too small to see easily with the naked eye. Various kinds of microscope, utilizing beams of light or electrons, are used to form magnified images of small specimens. Shapes and sizes of laboratory-grown crystals can be monitored (figs 61, 67), as can the way crystals might interlock to form a composite material (fig 58). Answers to industrial problems may require detailed information at the microscopic level. For example, it may be difficult to pump oil through sandstone if crystallizing minerals block its pores (fig 64).

In the scanning electron microscope (SEM), a fine beam of electrons is made to scan across the specimen. The electrons interact with the surface of the specimen and reveal the shapes present (figs 64, 66, 67). Very high magnifications can be achieved in this way (but electron images cannot reveal colour). In the light microscope, magnification is more limited, but colours in reflected or transmitted light can be seen. In addition, polarized light can be used to produce interference colours that provide information relating to the internal crystal structure (figs 60, 63, 65).

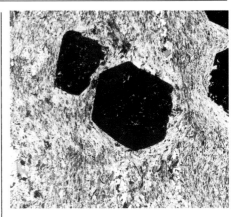

59   Garnet crystal shapes in a mica schist.

60   Granite: interlocking crystals, polarized light.

57   A salt crystal in a fluid inclusion in emerald.

58   Metal crystals in close-up: brass.

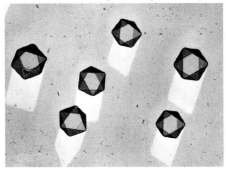

61   Silver bromide crystals in photographic film.

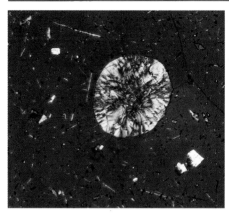

62 Crystals growing within volcanic glass.

63 Moon rock in polarized light, Apollo 11.

64 Quartz sandstone with blocked pores (SEM).

65 Agate: layers of minute quartz crystals. Interference colours of polarized light show individual crystals.

66 Granulated sugar (SEM).

67 Vitamin C crystals (SEM).

CRYSTALS AND LIGHT

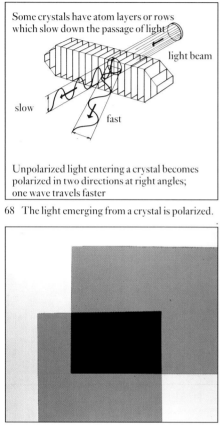

Some crystals have atom layers or rows which slow down the passage of light

light beam

slow

fast

Unpolarized light entering a crystal becomes polarized in two directions at right angles; one wave travels faster

68 The light emerging from a crystal is polarized.

69 Overlapping ('crossed') polarizing filters.

Light travels with equal speed in any direction through a transparent crystal with a cubic structure. But when ordinary light enters any non-cubic crystal it interacts with the atomic structure in a very interesting way: it is transmitted as two waves, travelling at different speeds. Each wave is polarized, that is it oscillates (or vibrates) in a single plane, and these two waves vibrate in planes at right angles to one another (fig 68). Much use is made of this optical property of crystals, especially in microscopes with polarizing filters (fig 69). A crystal placed between polarizing filters (with their vibration directions set at right angles – known as 'crossed') will appear a particular 'interference' colour that will depend on the crystal thickness and the relative speed of travel of the two light waves (figs 70, 71).

X-rays (light with a shorter wavelength than visible light) interact with crystals in other special ways – they are scattered by the electron clouds that define the positions of atoms in a crystal. Scattered X-ray waves recombine in patterns related to atom positions. This is known as X-ray diffraction, and is the main method used to work out the structures of crystals (fig 72).

70 'Interference' colours from polarizing filters.

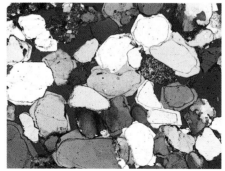

71 Sandstone between 'crossed' polarizers.

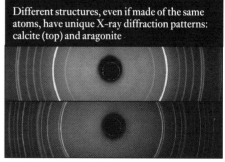

Different structures, even if made of the same atoms, have unique X-ray diffraction patterns: calcite (top) and aragonite

72 X-ray diffraction is used to identify crystals.

73 Light and crystals interact. A clear crystal of calcite gives a doubled image.

Double refraction

Light rays pass through transparent non-crystalline substances such as glass and through crystals of cubic structures with equal speed in all directions. But when unpolarized light enters a non-cubic crystal, such as calcite (fig 73), then the light travels as two component rays, polarized at right angles to each other. These two rays travel through the crystal with different speeds and in different directions. The two rays emerging from the crystal each give us a separate image – a double image is seen. Calcite shows this effect most distinctly, because of the particular arrangement of its atoms. Atoms in calcite occur in well-defined layers (fig 74). These layers do not offer much resistance to the passage of light waves vibrating at an angle to the layers, but light waves vibrating parallel to the atom layers are slowed down on their passage through the crystal. This is because light waves vibrating parallel to the layers interact strongly with the electron orbital clouds of the carbon and oxygen atoms of the carbonate groups.

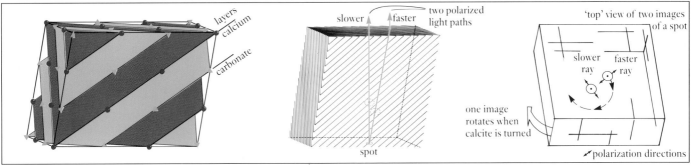

74 The layered structure of calcite affects the passage of light.

DEFECTS

Different kinds of defect occur in crystal structures, and these affect the physical properties of crystals. Some defects or imperfections, such as twinning (fig 78) and trapping of fluids (fig 80), have already been mentioned (page 6). Sometimes two different crystals grow interlocked on a very fine scale, and are often crystallographically aligned one to the other. For example, in 'star' rubies and sapphires (fig 76), light is reflected from inclusions of needles (usually rutile) aligned along three specific directions governed by the trigonal symmetry of the host mineral.

Other defects include vacancies, interstitial and substitutional atoms, and dislocations (fig 75). Under stress, atoms migrate by rearranging crystal defects – this is how rocks and glaciers move. When a metal bends, crystal dislocations are rearranging their positions atom-by-atom. Tiny iron carbide crystals in steel inhibit the movement of dislocations, making the metal tougher and difficult to bend (fig 81). Crystal defects are sometimes the cause of colour. For example, iron atoms substituting for silicon in quartz produce the purple colour of amethyst, while electrons trapped in vacant atom sites in sodium chloride (salt) cause a blue colour (fig 77).

76 'Star' gems: reflections from needles inside.

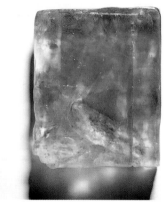

77 Colour caused by defects in a salt crystal.

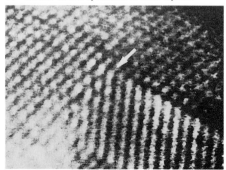

78 Twin defect (magnified 13 million times).

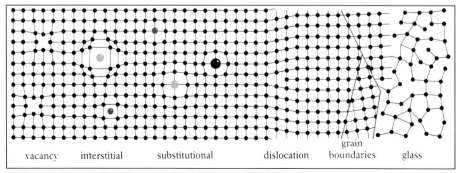

vacancy   interstitial   substitutional   dislocation   grain boundaries   glass

75 Some defect types found in crystals. Glass is not crystalline because its atomic structure is full of defects.

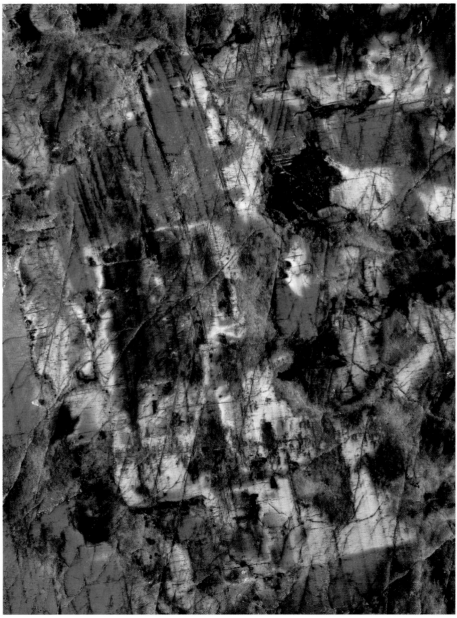

79 Iridescent colours produced by reflection of light from fine-scale intergrowths in labradorite feldspar.

Iridescence in feldspar
Labradorite feldspar can display multi-coloured iridescence (fig 79), caused by compositional defects. On cooling from high temperatures its aluminium, silicon, sodium and calcium atoms re-distribute, producing compositional fluctuations throughout the crystal on a very fine scale – these cause light to be multiply reflected. The colours we see are produced when the constituent waves (colours) of white light interfere and recombine after being reflected in this way.

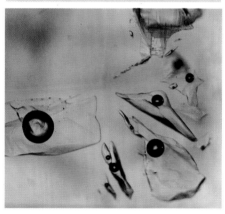

80 Gas bubbles and fluid trapped in fluorite.

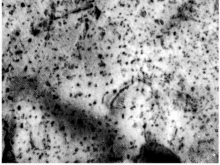

81 Tiny iron carbide crystals make steel tough.

HOW CRYSTALS GROW

For a crystal to start growing, a small number of atoms must first bond together to form a cluster – the nucleus of a new crystal. The ordered arrangement taken up by the atoms in this nucleation event will determine the eventual structure of the crystal. The crystal grows in size as atoms add onto the initial cluster one-by-one. Atoms are usually supplied (transported) to the growing crystal either as ions or as groups of atoms carried in a fluid, a melt or a gas. Some crystals can also grow within other crystals by atomic re-distribution (diffusion), nucleation and growth in the solid state.

Defects on the faces of crystals aid the growing process, as these provide sites for atoms to bond most easily onto the crystal. Although crystal faces may appear flat to the naked eye, high magnification reveals that they usually grow in a spiral fashion as a series of shallow ledges centred around surface defects (figs 84, 85). An orderly atomic structure grows outwards until the supply of atoms stops. The growth zoning shown by some crystals (e.g. fig 101) is evidence of interruptions that occurred in the supply of atoms, as well as changes in the type of atoms supplied.

82 Menthol crystals grown from vapour.

83 Alum crystals grown from water solution.

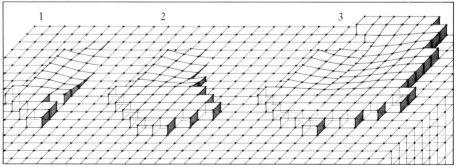

84 Spiral growth from a screw defect: unit cells of atomic structure build up in a spiral fashion.

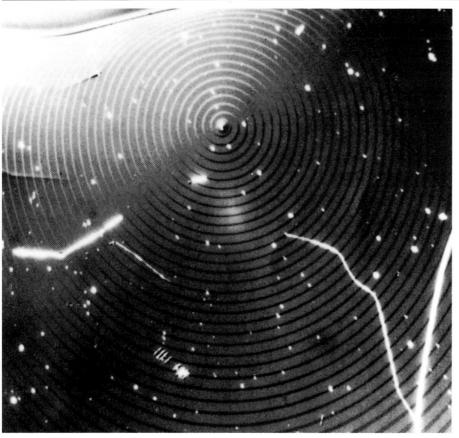

85 Spiral growth on a crystal face.

How to grow crystals

Alum (potassium aluminium sulphate) crystals (fig 83) can easily be grown in a jam jar. Dissolve powdered alum in warm water, stirring until no more will dissolve, then pour the solution into a jar. Cool, then suspend a single small crystal fragment of alum tied to a fine thread in the solution (fig 86). Leave the jar open for the solution to evaporate. Over the next few days atoms from the solution will add onto the 'seed' crystal, and it will grow in size. You can also grow crystals in a similar way from other common chemicals such as Glauber's salt, common salt, copper sulphate, citric acid and borax. To grow crystals from vapour, place menthol crystals in a small clear jar. Hold the outside of the jar in hot water until the crystals melt (at 44°C), then run the liquid around the inside of the jar. Watch crystals grow from the liquid as it cools. Seal the jar and keep it in a warm place. Needle-shaped crystals will continue to grow from the vapour over the following weeks and years (figs 82, 86).

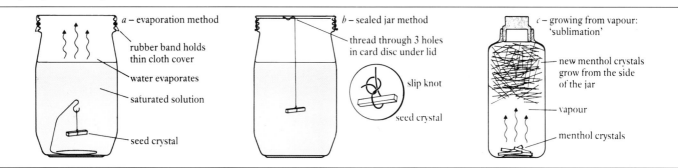

a – evaporation method
rubber band holds thin cloth cover
water evaporates
saturated solution
seed crystal

b – sealed jar method
thread through 3 holes in card disc under lid
slip knot
seed crystal

c – growing from vapour: 'sublimation'
new menthol crystals grow from the side of the jar
vapour
menthol crystals

86 How to grow your own crystals.

GROWING IN LIQUID

Most crystals grow from liquid solutions. The liquid is the carrier of the atoms or ions necessary for the growth of the crystal, and may be water or a molten substance at a high temperature.

Chemical substances may dissolve in a particular liquid – the solvent. The same solid substance may separate (crystallize) from the solvent when the temperature drops or when the liquid becomes saturated with the dissolved substance. The 'carrier' liquid may be water, a molten salt, a molten metal, a molten rock (magma) or an organic solvent.

Crystals grow from liquid in response to changes in either temperature, pressure or liquid composition. Many individual crystals may grow simultaneously if the nucleation rate is high (figs 89, 91). Molten rock is a complex mixture of atoms, and on cooling the solid that crystallizes will be a complex mixture of mineral crystals. Pure molten substances crystallize directly as they cool though their freezing point. For example, pure water crystallizes at 0°C (fig 87), while sapphire crystallizes at 2035°C (fig 88).

88 Crystals of synthetic ruby and sapphire.

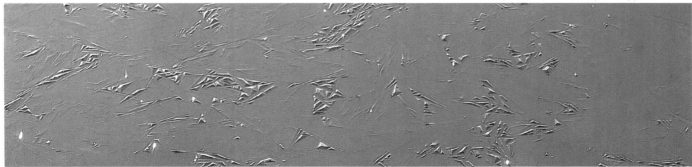

87 Ice crystals growing from water at the surface of a pond.

89 Crystals of saccharin grown from solution.

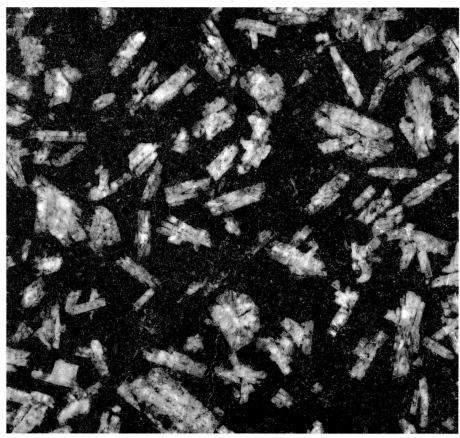

90 Synthetic quartz grown from hot water solution.

91 Crystals of feldspar which grew in molten rock.

Solutions and crystals

Solutions are formed when **solute** atoms, ions and small groups of atoms are dispersed – dissolved – within a liquid 'carrier' or solvent. Atoms of the liquid itself form transient bonds with solute atoms and 'hold' them in solution. The whole system is in motion, with bonds between solvent and solute continuously breaking and reforming. A liquid will dissolve a substance up to a certain limit at which point it becomes a saturated solution. Crystallization begins to occur if the temperature of a saturated solution drops, because then fewer solute atoms can be held in solution. However, small atom clusters or nuclei are a prerequisite for crystals to grow, and nuclei may sometimes have difficulty in forming. In such cases, the saturation point may be slightly exceeded and more dissolved substance may exist in solution making it supersaturated. For crystals to grow, a triggering event may be needed, such as the influx of 'seed' crystals to act as nuclei; crystals then grow until the solution is just saturated. In nature, minerals such as calcite, fluorite, quartz, silver and gold crystallize from water-rich fluids. Silicate minerals that crystallize from melts to form igneous rocks include olivine, feldspar (fig 91), mica and quartz.

GROWING IN SOLID AND VAPOUR

Crystals can nucleate and grow within other solid crystals. As atoms re-distribute in response to changes in temperature and pressure, they may take up different positions relative to one another, forming a new structural pattern. In this way one crystal structure (a phase) can be transformed into another. Crystals with different arrangements of the same atoms are known as polymorphs.

Sometimes the atoms of several different crystal phases become rearranged in the solid state to form an entirely new crystal phase (or phases) – this is a chemical and structural reaction – a process that happens continuously (but at different rates) within the Earth's rocks as they become heated and compressed at depth. The clay minerals in a shale can be transformed (over time at high pressure and temperature) into garnet crystals (fig 93). Likewise, the tiny calcite crystals in limestone can be converted into larger, but fewer, crystals of calcite that form marble (fig 94). Rocks that have been transformed by recrystallization into new mineral assemblages are known as metamorphic rocks. Some substances form crystalline solids directly from vapour without any liquid being formed (figs 92, 95).

92 Ice crystals grow as frost directly from water vapour.

93 Shale (left) slowly transforms into garnet–mica rock when compressed and heated within the Earth.

94 Heat and pressure convert limestone (left) into marble.

95 Yellow sulphur crystals growing from vapour escaping from cracks in the side of a volcano.

Phase transformations

When subjected to high pressures and high temperatures, crystals, and composite crystalline materials such as rocks (intergrowths of different mineral crystals), alter in various ways. In response to increased pressure the atoms in a crystal are forced closer together such that the volume they occupy becomes less. This compression can be monitored in the laboratory, using X-rays to measure the distances between atoms. As pressure increases still further, a phase transformation may occur as the atoms rearrange into a new more densely packed structure. Phase transformations account for the different density layers within the Earth and other planets.

An increase in temperature will cause an increase in the thermal vibrations of the atoms in a crystal and this will lead to an expansion of the (same) structure; at some point the pattern of atoms may rearrange (transform) into a new crystal structure more stable at high temperature.

For an increase in temperature and pressure simultaneously, there will be an interplay between the expansion and compression behaviour of the atomic structure and this will govern the way that the atoms are able to rearrange to bring about a phase transition.

EVIDENCE OF CRYSTAL GROWTH

96 Quartz with earlier growth shape enclosed.

Crystals often retain features from earlier stages of their growth. Zones of colour or inclusions of other crystals may outline the shape of previous growth surfaces (earlier crystal faces). Natural crystals frequently grow in a number of successive stages rather than in one steady episode. Previous crystal shapes may be visible inside clear crystals such as quartz (fig 96), where tiny pockets of fluid (from which the crystal grew) became trapped as the crystal resumed growth.

Crystal growth stages may also show up as colour zones, each with a slightly different chemical composition (fig 101); these are evidence that different types of atoms were being supplied by the fluid to the growing crystal at different times. A supply of chemically different atoms may alter the relative growth rates of different faces (page 4) and the crystal will then grow into a different shape (fig 101).

Sometimes crystals show evidence that they have begun to dissolve rather than grow (fig 97). This can happen if the crystal is exposed to a fluid or melt of different chemistry, or if the temperature and pressure conditions change.

97 Dissolution pits on a beryl crystal – their shapes and orientations are controlled by the symmetry of the crystal structure.

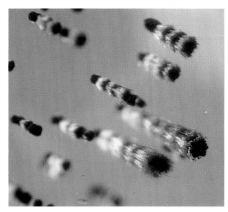

98 Crystal tufts of iron oxide within amethyst.

99 Growth architecture on a galena crystal face.

100 Diamond growth marks under the microscope.

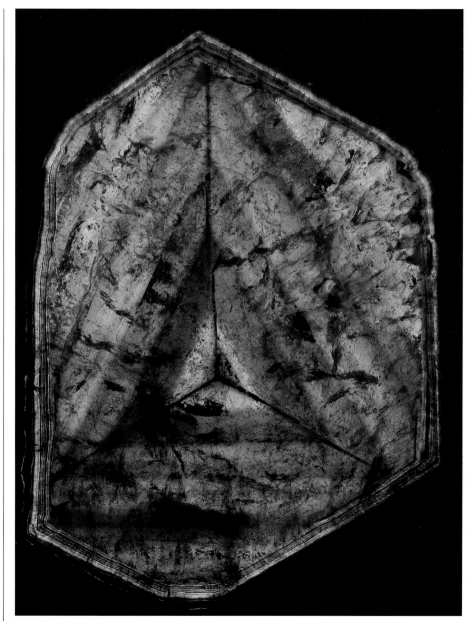

101 Colour zones reveal growth stages in tourmaline (a thin slice cut perpendicular to the crystal length).

GROWING CRYSTALS FOR INDUSTRY

Many different crystals are grown under laboratory conditions for a wide range of uses in industry. If required, crystals of very high chemical purity with structures almost free from defects can be synthesized by carefully controlling their growth conditions. Crystals required to react or dissolve quickly for certain applications can be grown under conditions that produce either very small crystals or ones with high surface areas (fig 103).

The electronic properties of some materials, such as silicon, can be adjusted by 'doping' pure crystals with tiny amounts of other elements. Crystals for optical devices such as lasers (fig 105) are grown with particular dopant chemistries to give special properties for different uses.

High quality quartz crystals, for use in electronic timing devices (quartz watches, see fig 107) are grown from hot water solutions (fig 90). Diamonds, for use as abrasives in special cutting and polishing tools, are grown at high pressure. Although not quite as hard as diamond, carborundum (figs 102, 104) is another important industrial abrasive, made by heating quartz, carbon, sawdust and salt in electric furnaces.

103   Salt: shapes grown for specific uses.

102   Carborundum (silicon carbide) crystals in abrasive grit.

104   Carborundum viewed in polarized light.

105 Crystals of yttrium aluminium garnet (YAG) doped with different lanthanide elements, for lasers.

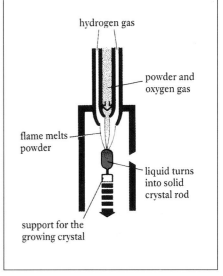

106 Growing synthetic crystals (above) in a Verneuil flame-fusion furnace.

In the diagram:
- hydrogen gas
- powder and oxygen gas
- flame melts powder
- liquid turns into solid crystal rod
- support for the growing crystal

Laboratory-grown crystals

In the Verneuil flame-fusion process (fig 106), a powder is rapidly melted in an oxygen–hydrogen flame, and the melt deposited onto the surface of a growing crystal held at a high temperature. The liquid crystallizes, adding to the growth of the crystal, which is slowly pulled away from the melt source in the furnace to produce a long single crystal. The exterior shape of a crystal produced in this way can look somewhat rounded (figs 88, 105), but inside it is one large single crystal with constant atomic pattern throughout. The quality and near-perfect structures of these crystals make them suitable for many special applications, such as the manufacture of crystalline materials used to produce laser beams, for making bearings in precision engineering, and for cutting into gemstones. Crystals of many different compositions are grown in this way, and can be 'doped' with controlled small amounts of impurities to give them special properties (fig 105). High quality single crystals can also be grown from solution in a molten 'flux' of metal oxides. Synthetic emerald (chromium-doped beryllium aluminium silicate) is grown in this way. Diamonds can be synthesized at very high pressures and temperatures by converting carbon into a cubic structure.

WORKING WITH CRYSTALS

107 The quartz crystal used as a time-keeper (inset) in a watch.

A knowledge of the inner structures and properties of crystals means that we can exploit crystals in beneficial ways.

Quartz has a crystal structure that is piezoelectric – it exhibits the property of pressure-induced electricity, as used in gas cooker lighters. If a quartz crystal is squeezed, then its structure deforms, and opposite charges appear temporarily on opposite sides of the crystal. This property also means that when electrical charges are applied to each side of a slice of quartz, its structure expands and contracts with a specific vibration frequency. Quartz crystals are used as time-keepers in quartz watches to accurately control the oscillation frequency of an electrical circuit and thereby keep the timing of an alternating electrical charge constant (fig 107).

Winter temperatures can mean trouble for diesel engines: wax crystallizes in diesel fuel, blocks fuel filters and engines stop working. The hydrocarbon chain molecules in diesel fuel take up an ordered arrangement and bond to each other at low temperatures forming large plate-shaped crystals – just a few of these block the filter. Molecular additives have been designed to restrict the size and modify the shape of these plate-like wax crystals, so that a slurry of smaller crystals can pass through the filter to the warmer parts of the system where they re-dissolve, and the engine keeps on working (fig 109).

Chocolate contains crystals formed of cocoa butter molecules, but these can form different crystal structures (polymorphs) in chocolate. Manufacturing a good quality, crisp-breaking, smooth-tasting chocolate means controlling the proportions of these different polymorphs that crystallize as chocolate cools (fig 110).

Clays, pottery and ceramics

Since earliest times clay has been worked, moulded and fired to produce pots. This is possible because of the physical properties of clay crystals. Clay minerals have atomic structures consisting of thin aluminosilicate layers that form flat, flaky, plate-like crystals (fig 108, inset). These crystals are very small, and surface charges hold them together by electrostatic attraction, making them into a cohesive mass.

Water molecules are likewise attracted and surround each clay crystal, helping to bind them together and impart plasticity to the bulk clay. The surrounding water layers allow the clay crystals to slip past one another, so clay is easily moulded when wet (fig 108). Kaolinite is the most widely used clay mineral in making pots. Firing drives off the water, and the clay transforms into new harder crystal structures at high temperatures (page 35), resulting in a durable product. Kaolinite, the clay mineral in china clay, is used to make high quality porcelain and insulator ceramics.

108 Making pots relies on the cohesive and plastic properties of clay.

109 We can prevent fuels freezing because we know their crystal structures.

110 The right crystals make chocolate taste good.

SAME ATOMS, DIFFERENT CRYSTALS!

Sometimes there is more than one way to arrange particular atoms to form a crystal and these different forms, or polymorphs (page 34), can have very different appearances and uses.

A good example of this is the element carbon. Until recently there were two known forms of solid carbon: diamond and graphite (figs 111–115). The carbon atoms in graphite are arranged in layers (fig 114) which can slip past each other easily, so graphite is very soft and is used in pencils and as a lubricant. In diamond, the carbon atoms are strongly bound to each other in a highly symmetrical arrangement in three dimensions (fig 114), so it is very hard and used as an abrasive, as well as being the most prized of gems due to its beauty and durability (fig 113). A third solid form of carbon has recently been found. The famous C_{60} 'buckyball' form (fig 117) has led to the production of many similar structures, where graphite layers are rolled up into balls or tubes (fig 116).

Polymorphism is also found in more complex chemicals, including some pharmaceuticals, whose polymorphic form can affect how much of it is absorbed by the human body, or how long a medicine stays stable and safe.

113   Close-up of the Koh-i-Noor diamond.

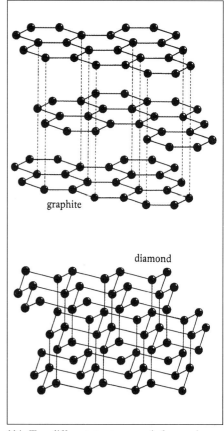

graphite

diamond

114   Two different structures made from carbon.

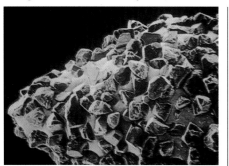

111   Diamond abrasive grit in your dentist's drill.

112   Hard diamond contrasts with soft graphite.

SAME ATOMS, DIFFERENT CRYSTALS!

115   Diamond gem surrounded by crystals.

Buckyballs

Three scientists, Professors Curl, Kroto and Smalley, were awarded the 1996 Nobel Prize for Chemistry for their discovery and synthesis of the novel C_{60} form of carbon, also known as 'buckminsterfullerene', 'fullerene' or simply 'buckyballs' (fig 117). Shaped just like a soccer ball, this prototype fullerene has since been studied extensively by scientists, who have produced many different derivative forms and compounds of it.

Among the derivatives have been other shapes, made by adding further carbon atoms to the basic fullerene and twisting it round to form different frameworks – like a sort of molecular chicken wire! The most interesting of these is perhaps the 'buckytube' or carbon nanotube (fig 116). It is thought that structures like these may have applications in chemical catalysis, as tiny factories where special chemical reactions can take place.

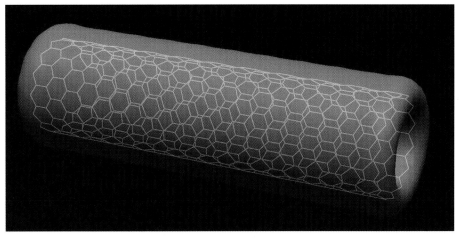

116   Layers of carbon atoms – as in graphite – rolled up into a tube shape, forming a 'nanostructure' called a buckytube.

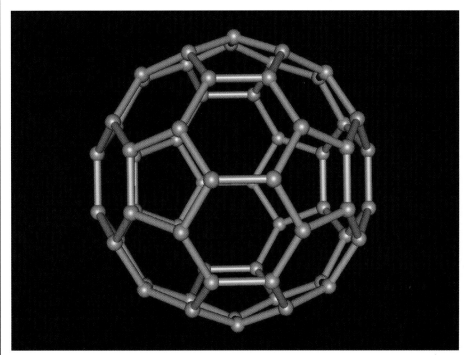

117   The famous C_{60} buckyball molecule.

ELECTRONICS AND LIGHT

Crystals can be found everywhere in the worlds of electronics and optics. These industries make use of the special properties that arise from having certain types of atoms and molecules bound closely together in an ordered fashion – or, as previously mentioned (see page 28), from what happens when that order is disrupted somehow.

Materials used in electronics get their properties from the way in which an electrical current can pass through a crystal structure of certain kinds of atoms. Silicon is a good example – its electronic properties can be controlled by 'doping' pure crystals with tiny amounts of other elements, during the very carefully controlled growth process.

Laser beams (fig 118, 120) can be produced because of the ordered nature of crystals. Certain crystals, such as KTP (see box), can be used to change the frequency and wavelength (hence the colour) of the light, to tailor it to a particular use, for example in a CD player, for medical research and surgery, in fibre-optical communications, or simply for a laser show.

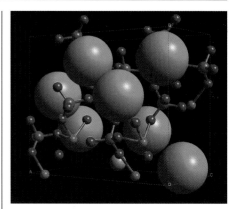

119   A crystal structure called KTP, with special optical properties.

KTP

KTP (fig 119) is potassium titanyl phosphate ($KTiOPO_4$). It has 'non-linear' optical properties because of its crystal structure. If light of a particular frequency (e.g. laser light) passes through KTP in a certain direction, then the frequency emitted is doubled. Such crystals can therefore be used to provide some tunability, or selection, for the wavelength of light obtained from a laser.

118   The red beam is from a type of laser particularly useful in medicine.

120 A researcher working with a laser.

121 CVD crystal growing rig.

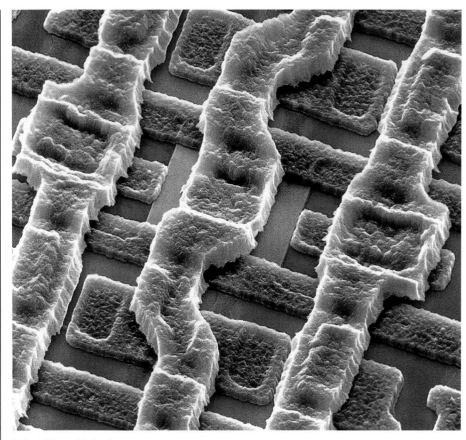

122 A silicon chip in close-up.

Chemical vapour deposition (CVD)

Chemical vapour deposition is the most common way of growing thin crystalline layers for important applications in the electronics industry, e.g. the production of solid-state devices such as microprocessors (fig 122). In a CVD process, the compounds to be deposited – or a mixture of chemical reactants which will together produce the compounds – flow in the gaseous phase over the 'substrate' to be coated (fig 121). In the case of electronic devices, the substrate is often pure silicon. Conditions at the substrate's surface are carefully controlled so as to promote the growth of the required layers.

Factors that can be varied include temperature, composition of the gas mixture, and flow rate. Sometimes, chemical reactions in the vapour are induced by a flash of laser light or an electron beam. Thus the thickness and composition of the crystalline layers can be tailored for the application required. For electronic devices, many layers of different compositions are usually deposited. These may include silicon doped with impurities (which change the electronic behaviour), gallium arsenide, or indium phosphide. The CVD process may be carried out several times, interspersed with etching procedures whereby some of a top layer is removed, following the addition of a template, to produce the final electronic component.

LIQUID CRYSTALS

Almost every electronic device now produced contains a liquid crystal display (LCD). But the term 'liquid crystal' seems contradictory! Crystals are ordered arrays of atoms and molecules, whereas liquids are random collections of molecules or atoms in motion.

Liquid crystals (figs 123, 127) fall somewhere between these two extremes. They consist of molecules that are arranged in approximately ordered fashions in *some* directions, but not in all three dimensions (fig 126). The arrangement is usually due to some combination of molecular properties, such as shape and the difference in electronic charge (positive and negative) between regions of the molecule, and can easily be altered by environmental factors such as temperature or external electric field.

LCD devices work because when a tiny electric field is applied the molecules line up to let polarized light through in one direction only. The field is applied to parts of a display only, allowing different characters to be seen by the viewer (fig 124). LCDs are very energy efficient, making minimal demands on electricity sources such as batteries.

124 A liquid crystal seven-segment display.

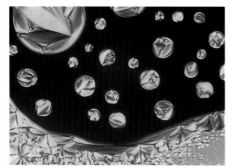

123 Liquid crystals growing.

The seven-segment display
This is perhaps the most familiar way in which liquid crystals are used. Liquid crystal seven-segment displays make up the digital readouts on watches and calculators, and are widely used for information panels on the front of household electronic devices as diverse as CD players, cookers and cameras.

The display is made up of seven segments in a 'figure of 8' arrangement.

A reservoir of liquid crystal solution is sandwiched between a transparent surface and the segments themselves on the back panel. In response to signals, a small electric field is applied through the solution to different combinations of segments, rotating the molecules in the liquid crystal and making the solution opaque in certain areas. The digits 0 to 9 can therefore be displayed as required.

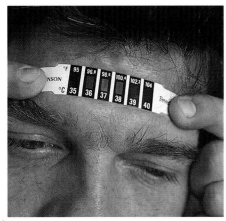

125 A thermochromic strip thermometer.

Thermochromic liquid crystals
Some liquid crystals are very sensitive to temperature changes. The angle of tilt of the molecules in the ordered regions of these crystals changes with temperature, and this in turn changes how the system reflects light. We see this as different colours. This effect has found uses in thermometers (fig 125), and in the dyes for those briefly fashionable T-shirts that changed colour with temperature!

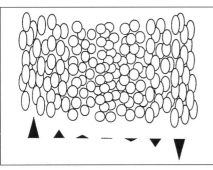

126 Liquid crystal structure.

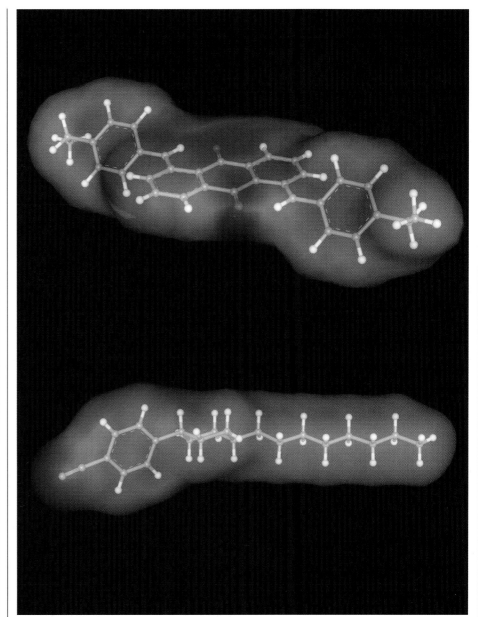

127 Two kinds of molecules that exhibit liquid crystal behaviour at around room temperature.

CRYSTALS AND YOU

128   Protein crystals (glucose isomerase).

129   A simulated Laue pattern for a protein.

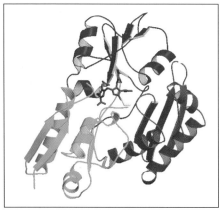

130   The molecular structure of an enzyme protein.

The beautiful pattern of spots on the facing page (fig 131) is a record produced by the interaction of X-rays with a tiny crystal of a protein – one of the chemical compounds out of which living things are made. Most of what we know about the atomic and molecular structures of crystals comes from investigation of this sort of 'diffraction' pattern. The intensities and positions of all the spots depend on the arrangements of atom and molecules inside the crystal, so by measuring the positions and intensities of thousands of spots very accurately, we can work back using powerful computational methods to produce 'pictures' of the molecules. You can see examples of these here (fig 130) and on other pages of this book (e.g. pages 47, 51, 53), shown in display styles designed by scientists to be easy to visualize various important features of different kinds of molecule and structure.

Detailed studies like this of crystals of all sorts of materials, ranging from rocks and minerals to the components of life itself, enable us to understand how things work in the world around us, and therefore how to control and design aspects of it in order to make our lives better. The car you ride home in, the clothes you wear, the medicines you take and even the foods you eat have in some way been improved and developed over many years by scientists studying crystals.

We hope this book is showing you that as well as being very beautiful, crystals are very important too. Up to now, most of the crystals we have discussed have been of non-living substances. If you read on, you will see how biological materials, including medicines, also owe much to crystals …

131 Protein Laue pattern.

THE RAW MATERIALS OF LIFE

A large part of our understanding of how living things function is due to the study of crystals. The materials that carry out the processes of life itself, and the compounds out of which living things are made, can also be formed into crystals and studied using X-rays.

Biological molecules are much larger and more complicated than those of chemicals such as water, alcohol, aspirin or even plastics. They are therefore more difficult to crystallize and study, and it is only in the last 30 to 40 years or so that it has been possible to determine the shapes of these molecule. And the secrets of how they work are locked up in those shapes …

These molecules are so important to us that scientists go to a great deal of time, trouble and expense in trying to crystallize and study them (figs 132–135). It can take months for a single, sub-millimetre-sized crystal of a protein to grow, under carefully controlled conditions of temperature and solution concentrations. Even then, although a crystal may look good optically, it may prove to be of poor quality when examined by X-rays. Scientists have even tried to grow crystals in space to see whether they grow more easily in the absence of gravity!

It is estimated that there are over 100,000 different biological molecules in the human body alone. With the huge variety of other organisms on this planet, made up of different molecules, the number of naturally occurring such molecules is truly immense! The structures of only about 10,000 such molecules have been reported by the end of the 20th century, so we have a very long way to go to understand all the details of our world at the molecular level.

132   Crystals of lysozyme.

133   Crystals of thaumatin.

134   Crustacyanin isolated from lobster shells.

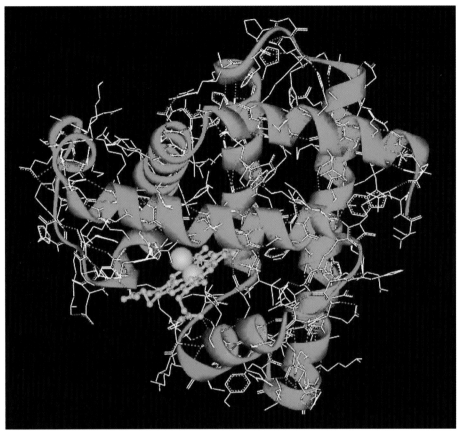

135   Haemaglobin – the protein molecules found in red blood cells.

Protein crystallography

In order to understand the function of biological molecules, to design new medicines or to engineer molecules with improved properties for industrial applications, one needs to know the structure of the molecule of interest. Structure is determined by X-ray diffraction, which can only be applied to ordered assemblies of molecules – i.e. crystals. Determining the structure of large biological molecules like proteins is challenging. The same mathematical methods used to derive the structures of smaller molecules from diffraction patterns also work for larger molecules, but it is often helpful to use starting models for the molecules, derived from either pure computational procedures or from a previously determined protein structure. Sometimes, special derivatives of the protein of interest are prepared which contain heavier atoms, and these help in the structure's determination. Protein crystals usually contain large amounts of water, so crystallographers need to consider how water molecules interact with the protein molecule as they solve its crystal structure.

136   Synchrotron Radiation Facility, France.

Synchrotron radiation

Although it is usually possible to study protein crystals using conventional X-rays, special X-ray sources called 'synchrotron sources' (fig 136) are often used to make these studies quicker and of higher quality. Synchrotron sources are huge (and extremely costly) circular machines, tens or even hundreds of metres across. The X-rays generated by synchrotrons are many times more intense than X-rays produced in ordinary laboratories, and have other special properties (such as the ability to choose any X-ray wavelength, and a very parallel nature) which make them particularly suitable for studying biological molecules. Governments, industrial companies and other research organizations worldwide are now recognizing the benefits that medical and pharmaceutical research can gain from synchrotron studies.

MAKING BETTER MEDICINES

Our knowledge of the shapes of biological molecules helps us to understand how they work – and this in turn gives us a way of influencing their function. For example, if a biological molecule is performing a function that is harmful to us, we can (if we know its shape) design a small molecule that can bind to it and prevent it from causing us harm.

Likewise, if we understand how known medicines bind to biological molecules, and how they help us, we can use that knowledge to design new molecules to do the job better, perhaps with a lower dose or fewer side-effects. In the same way that scientists study crystals of biological molecules, they can grow and study crystals of those molecules with the small drug molecules attached to them (fig 138).

This is one of the ways in which pharmaceutical companies and university researchers devise new treatments to keep you and I, or our pets and farm animals, healthy – all based on studies of crystals (figs 139–141).

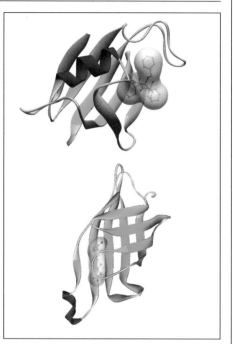

138 Two protein molecules with smaller molecules attached.

Molecular modelling
Molecular modellers (fig 137) use computational methods to help them analyse how biological, chemical and physical processes occur at a molecular level, and to help them come up with ways of improving or preventing such processes. The complexity of the systems being studied means that this field is always pushing the limits of computer graphics technology, mathematical theory and computational speed. There are a small number of commercial organizations who provide software for industry and academics.

137 A molecular modeller at work, using up-to-date computer technology.

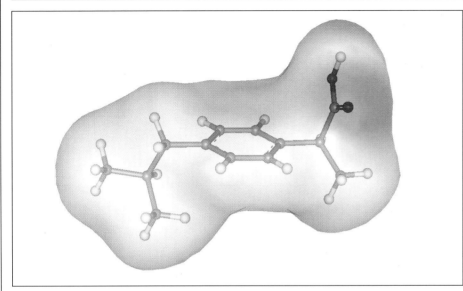

139   The molecule in the drug ibuprofen (to relieve headaches).

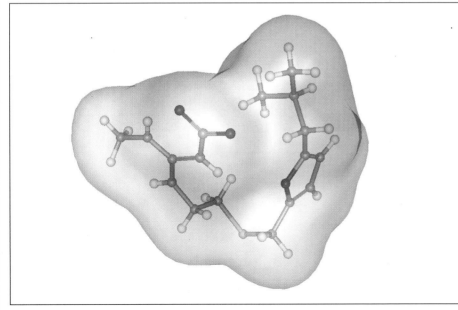

140   The molecule in the drug ranitidine (for relief of stomach pains and indigestion).

Drug development

The design of a new drug molecule, as described here, is just the first stage in producing a new medicine. The study of crystals also has a role to play in the rest of the molecule's journey from being a model on a computer to a medicine on a shelf.

Of course, the molecule has to be tested to determine whether it works in real life and whether it has harmful side effects ('toxicology'). A scheme for manufacturing the drug molecule on the appropriate scale, and a formulation (i.e. a pill, a solution, a skin patch etc.) in which to deliver the molecule in the best possible way, both have to be worked out. Before a pharmaceutical company is allowed to market a drug, regulatory authorities have to be satisfied that everything possible is known about the drug's behaviour under any circumstances imaginable. Crystallography plays a vital role in the research process.

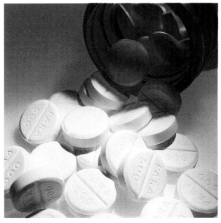

141   Crystal studies are vital in new drug design.

CRYSTALLINE LIFE

One step further still from crystallizing biological molecules is the crystallization of assemblies of biological molecules – living things themselves. Incredibly, it is possible, with a great deal of effort and patience, to produced ordered crystals of the smallest living things of all – viruses.

These crystals have been studied by the same techniques used to study proteins, and in some cases (the number of these is increasing all the time) it has been possible to determine where *all* the atoms are in a whole virus! This is helped by the fact that viruses are highly symmetrical and often regularly shaped. Such studies are helping us to understand virus life processes, in turn helping us to come up with ways of defeating their harmful effects on our lives and on those of animals and plants.

These incredible feats of science are increasing the scope of our understanding of the world around us from atomic levels up to the region where microscopes can see things. With the help of crystals, there is now no scale of structure, from metres down to the atomic, where we cannot probe.

The foot-and-mouth disease virus
One recent triumph of virus crystallography was the solution of the structure of the foot-and-mouth disease virus by a team from Oxford University in the UK. Foot-and-mouth disease (figs 142–145) is a serious viral infection of cattle – outbreaks of the disease have to be contained by slaughter of entire herds. The work on the virus structure had therefore to be carried out under the strictest possible quarantine precautions. With the use of the UK's Synchrotron Radiation Source (SRS, see page 51), many carefully prepared crystals of the virus and many months of intense and complex computational work, the virus' structure was determined down to the level of individual atoms and is shown here (fig 145). Notice how highly symmetrical it is – the structure displays 'icosahedral symmetry', which is common to many virus structures.

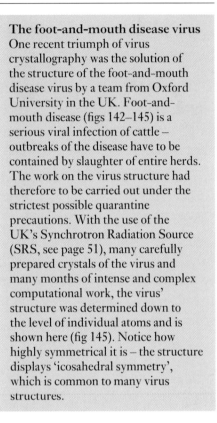

142 Healthy cattle.

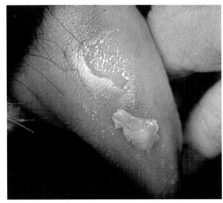

143 Foot-and-mouth virus infection.

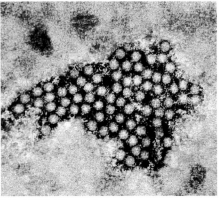

144 TEM of foot-and-mouth virus.

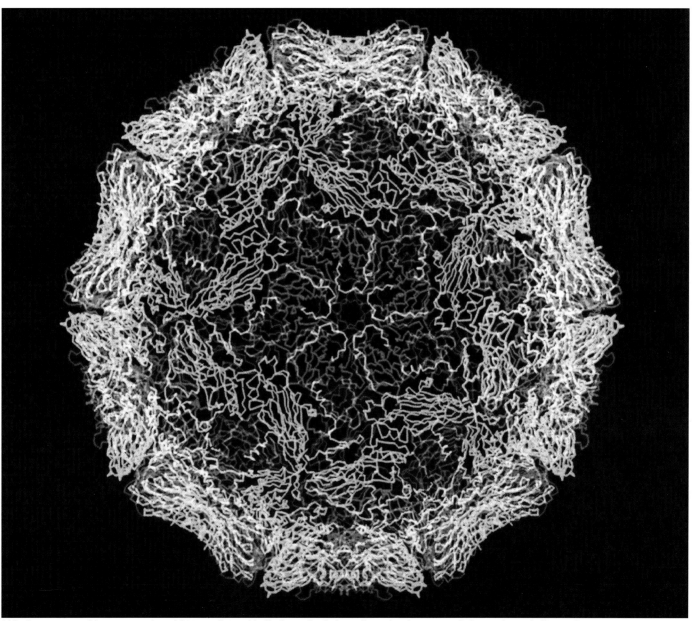

145 Computer graphic representation of the crystallographically determined, three-dimensional structure of a foot-and-mouth disease virus.

FACTS AND FIGURES

CRYSTALS AND CRYSTALLOGRAPHY

The word *crystal* is from Greek, κρύσταλλος *krystallos*, from *kryos:* icy cold. Pliny the Elder, Book 37, wrote, 'a violently contracting coldness forms the rock crystal in the same way as ice'. That rock crystal, clear quartz, is made from ice so deeply frozen that it will not thaw was an idea that existed in China, Japan and Alaska, and people believed this into the 17th century. Robert Boyle demonstrated that quartz is far too dense to be a form of ice. Solinus had reached a similar conclusion in the 3rd century BC.

Late 1660s: crystallography, the study of crystals, started in earnest. The crystal systems (see page 16) were established in the latter half of the 17th century after crystals had been studied in terms of external shape and the stacking of spheres.

1665: Robert Hooke, observing stacked musket balls, reasoned that alum crystal shape could be built up from minute spheres.

1669: Niels Steensen (Steno) established that the angle between equivalent faces on different quartz crystals is always the same, regardless of growth distortions.

1690: Christian Huygens, in his *Treatise on Optics,* explained the structure of Iceland spar (calcite - page 27) in terms of a 3-D pattern of 'squashed spheres'.

Late in the 18th century crystal structure was looked at in terms of minute 'building block' units.

1781: René Haüy accidentally broke a calcite crystal, and its orderly cleavage angles suggested a regular stacking of elementary building blocks. His crystal structure theory, published in 1784, was a great step in the science of crystallography.

Around 1780: Romé de l'Isle measured many crystal face angles, and the *Law of Constancy of Angle* was established: *in all crystals of the same substance, the angles between corresponding faces have a constant value.*

1830: J F C Hessel derived the 32 different 'point groups' of 3-D symmetry, also known as the crystal *classes* (see page 16).

1848: Auguste Bravais found that there are just 14 different types of arrangement of regularly repeating points in three dimensions: the 'Bravais lattices' (see page 3).

1885: E von Fedorov established the existence of 230 different sorts of 3-D patterns. These 'space groups' represent the different ways that groups of atoms 'clothe' the 14 'Bravais lattices' (see page 15).

1912: X-rays took the study of crystals deep into their atomic structure. Max von Laue reasoned that atomic patterns of crystals should help him establish the wavelength of X-rays. W Friedrich and P Knipping aimed an X-ray beam at copper sulphate crystals and obtained successful diffraction patterns.

1913: W H and W L Bragg turned the experiment around: using the newly found wavelength value, they used X-rays to determine the structures of various crystals using X-ray diffraction. The first crystals to have their structures analysed were potassium chloride and sodium chloride.

Crystal structures were now seen to be built up in repeating units of atomic pattern rather than from solid shapes.

An early wooden model octahedron showing small cube units as 'building blocks'.

Left: Clay model of pyrite made to illustrate the work of Romé de l'Isle for his *Crystallographie,* 1783. Right: Wooden model of calcite made by Count de Bournon, in the Greville Collection obtained by the Natural History Museum in 1810.

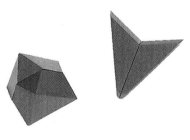

Bronze pivoted models to show twinning in gypsum (right) and spinel, made before 1811.

SYSTEMS AND SYMMETRY: CLASSES

The table below shows the 11 most common crystal classes (see page 16) found in minerals.

There are 32 classes in all. Some are rare. A typical mineral is named for each class. Classes are also called 'symmetry groups', or 'point groups'. The number of each kind of symmetry possessed by each class (see pages 14 and 15) is listed:

System name	Kinds of symmetry					
	Axes (x-fold)				Planes	Centre
	2	3	4	6		
CUBIC						
diamond	6	4	3	–	9	yes
pyrite	3	4	–	–	3	yes
tetrahedrite	3	4	–	–	6	no
TETRAGONAL						
zircon	4	–	1	–	5	yes
TRIGONAL						
quartz	3	1	–	–	–	no
calcite	3	1	–	–	3	yes
tourmaline	–	1	–	–	3	no
HEXAGONAL						
beryl	6	–	–	1	7	yes
ORTHORHOMBIC						
topaz	3	–	–	–	3	yes
MONOCLINIC						
orthoclase feldspar	1	–	–	–	1	yes
TRICLINIC						
plagioclase feldspar	–	–	–	–	–	yes

CRYSTALS IN SYSTEMS

CUBIC: galena, pyrite, garnet, diamond, spinel, halite (common salt), fluorite, zinc blende (sphalerite), alum, magnetite, tetrahedrite, cuprite, pitchblende, copper, gold, silver, iron, lead, platinum, silver bromide, quicklime (calcium oxide).

TETRAGONAL: cassiterite, zircon, wulfenite, scheelite, scapolite, idocrase, chalcopyrite, rutile, urea, potassium hydrogen phosphate.

TRIGONAL: tourmaline, corundum (ruby and sapphire), quartz, calcite (incl. coral, stalagmite), dolomite, magnesite, hematite, dioptase, cinnabar, Chile saltpetre (sodium nitrate), carborundum (silicon carbide).

HEXAGONAL: ice, graphite, apatite, beryl (incl. emerald and aquamarine), nepheline, zinc, iodoform, silver iodide, menthol.

ORTHORHOMBIC: baryte, topaz, olivine (peridot), nitre (saltpetre, potassium nitrate), staurolite, chrysoberyl, epsomite (Epsom salts), aragonite, alpha-sulphur, cordierite (iolite), zoisite (incl. tanzanite), stibnite, anhydrite, marcasite, natrolite, Rochelle salt, codeine, strychnine, atropine, oxalic acid, citric acid.

MONOCLINIC: orthoclase feldspar (incl. moonstone), gypsum (selenite), mica, jade (jadeite and nephrite), steatite (soapstone, talc), asbestos, malachite, chlorite, epidote, amphibole (incl. hornblende), pyroxene (incl. augite), serpentine, clay minerals (including kaolinite), spodumene, borax, beta-sulphur sucrose (and other sugars), tartaric acid, washing soda (sodium carbonate decahydrate), baking powder (sodium bicarbonate), naphthalene, Glauber's salt.

TRICLINIC: turquoise, plagioclase feldspar (incl. labradorite and sunstone), rhodonite, axinite, kyanite, copper sulphate.

Part of a collection of 24 cut-glass models of the crystal forms of precious stones in natural colours; M. de Struve 1825.

Late 19th century wooden model hexoctahedron, one of a set of over 700 accurate pearwood models made by the firm of Krantz in Bonn.

Rhombohedron in glass sheet, with thread axes; from a set made to order by Dr Krantz in Bonn, 1896–1901.

FACTS AND FIGURES

THE WORLD'S LARGEST CRYSTALS

THE WORLD'S LARGEST AUTHENTICATED CRYSTAL

A beryl crystal, Malagasy Republic.
18 metres long, 3.5 metres in diameter,
weight: about 380 tonnes.

Other large mineral crystals:

Mineral		Size (m)	Weight (kg)
fluorite	New Mexico	2.13	16,000
calcite	Iceland	7 × 7 × 2	254,000
garnet	Norway	2.3	37,500
topaz	Mozambique	0.91	2,677
mica	India	4.57 × 3.05	77,000

• A quarry in Colorado, USA, might have been in a single crystal of microcline feldspar. Its size would have been 49.38 × 35.97 × 13.72 metres, with an estimated weight of about 16,000 tonnes.

• The largest known crystal from space is a single iron–nickel alloy crystal in a meteorite found in the former USSR. It measures 920 × 540 × 230 mm and weighs 303 kg.

• The largest diamond crystal known was the Cullinan found in 1905 in South Africa. It weighed 3106 carats (about 621 g).

THE LARGEST FACETED CRYSTALS

• The largest cut diamond is the Cullinan I, 530.20 carats, in the British Crown Jewels.

• The heaviest faceted gem is a yellow topaz, at 22,898 carats (over 4.5 kg).

• The largest cut gem is a light yellow 'champagne' topaz of 36,854 carats weighing 7.37 kg.

OTHER RECORDS

• The longest known stalactite (from the cave roof) is 7 m long (The Poll Cave, County Clare, Ireland).

• The longest stalagmite (from the cave floor) is 29 m long (Aven Armand Cave, France).

• The longest complete column is 39 m long (Nine Dragons Cave, China).

• The smallest brilliant cut diamond is 0.22 mm in diameter, weighing 0.0012 carats (0.00024 g).

INTERATOMIC DISTANCES
in nanometres (nm)
(one nanometre = a millionth of a millimetre)

In *quartz* and other silicate minerals (see pages 8–11) silicon to oxygen distance ranges from 0.152 to 0.160 nm.

In *ice* the distance between oxygen atoms is 0.26 to 0.27 nm (a very 'open' structure).

In *diamond* the carbon atoms are 0.154 nm apart. In *graphite* they are 0.145 nm apart (in parallel planes) and 0.335 nm apart (across the planes; see page 42).

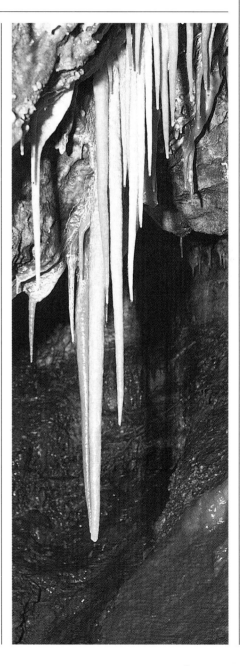

Stalactites growing from the roof of Treak Cliff Cavern, Castleton, Derbyshire, UK.

INDEX AND CREDITS

PICTURE CREDITS
Images from The Natural History Museum, London unless otherwise indicated.

Front cover, Figs 128, 132, 133, 134 Naomi Chayen, Imperial College, London
Figs 1, 26, 28, 47, 48, 65, 66, 67, 78, 87, 89, 90, 92, 104, 111, 122, 123 Science Photo Library
Figs 6, 98 Cassandra Goad
Figs 27, 64 British Geological Survey
Fig 32 Katia and Maurice Krafft
Fig 40 Christine Woodward
Fig 45 Spink & Son Ltd.
Fig 46(inset) Eric Robinson
Fig 51 Ian F Mercer
Fig 57 E. Alan Jobbins
Fig 61 Kodak Limited, Research Division/ Journal of Photographic Science
Fig 85 Elisabeth Bauser
Fig 100 Roy Huddlestone
Fig 103 ICI Chemicals and Polymers
Fig 105 Crystalox Ltd.
Fig 107 Louis Newmark plc/Swatch
Fig 108(inset) English China Clays plc
Fig 108 Josiah Wedgwood & Sons Ltd.
Fig 109 Quadrant
Fig 113 Crown copyright: Historic Royal Palaces
Figs 116, 117, 127, 135, 138, 139, 140 Cambridge Crystallographic Data Centre
Fig 118 Lawrence Livermore National Laboratory/Science Photo Library
Fig 119 Molecular Simulations Inc., Cambridge, UK
Fig 120 Hank Morgan/Science Photo Library
Fig 121 Dept. of Chemistry, Queen Mary and Westfield College
Fig 125 Francoise Sauze/Science Photo Library
Figs 129, 130 Reproduced with the permission of Dr A Haedener and Professor J R Helliwell as well as the International Union of Crystallography.
Figs 131, 141 Daresbury Laboratory
Fig 136 Artechnique/European Synchrotron Radiation Facility
Fig 137 Glaxo Wellcome
Figs 142, 143, 145 Professor D. Stuart and Dr E. Fry, Department of Molecular Biophysics, Oxford University
Fig 144 Andrew Syred/Science Photo Library

ACKNOWLEDGEMENTS
With thanks to Dr Stephen Maginn, Cambridge Crystallographic Data Centre and the British Crystallographic Association, for his contribution on the application of crystals in industry and supply of associated images.

First edition designed by David Robinson
Second edition designed by Peter Dolton
Typesetting by Cambridge Photosetting Services
Colour reproduction by Atlas Mediacom Pte Ltd., Singapore
Printed by Vallardi Industrie Grafiche S.p.A., Milan, Italy

Reprinted for The Stationery Office, 1994

ISBN 0 565 09145 X

A catalogue record for this book is available from the British Library

Metric conversions

gram	= 0.035 ounces
tonne	= 1.1 ton
metre	= 39.37 inches
kilometre	= 0.62 miles
0°C	= 32°F

FURTHER READING

Alan Holden and Phylis Morrison
Crystals and Crystal Growing
MIT Press
London; Cambridge, Mass. 1982
318pp ISBN 0-262-58050-0
The essential handbook for crystal-growers,
including several recipes and alternative growing
methods – with a great bonus in the clear
explanations of crystal optics, structures and
symmetry and plenty of good diagrams and
photographs.

Christine Woodward and Roger Harding
Gemstones
The Natural History Museum; London 1988
60pp ISBN 0-565-01011-5
Packed with full-colour photographs and diagrams,
this guide provides a start into the world of gems
and gem minerals, their origins, descriptions, fakes,
internal surprises and fascinating properties.

R F Symes
Rock and Mineral
Dorling Kindersley/The Natural History
Museum; London 1988
64pp ISBN 0-86318-273-9
Brilliant colour photographic guide to the wide
range of mineral-related subjects with information
on origins of the different rocks, on crystals and
their shapes, precious metals, gemstones, pebbles,
pigments, caves, useful minerals, etc.

Deirdre Janson-Smith with Gordon Cressey
Earth's Restless Surface
The Natural History Museum; London 1996
60pp ISBN 0-11-310056-6
This book is all about change at the Earth's surface,
exploring the dynamic processes and evidence of
change. Illustrated in colour throughout.

Elizabeth A Wood
Crystals and Light
Dover Publications; New York.
2nd revised edn. Constable; London 1977
160pp ISBN 0-486-23431-2
The clearest beginners' guide to the interaction of
light and crystal structures. Includes plenty of
detailed, advanced information on structure,
symmetry, optics and optical microscopy, all in a
readable style with many diagrams.

M H Battey
Mineralogy for Students
Longman; 2nd edn. London 1981
355pp ISBN 0-582-44005-X
A clearly written book with good diagrams for those
who need advanced information and detailed
descriptions of crystals, minerals and optical
microscopy methods.

Rodney Cotterill
*The Cambridge Guide to the
Material World*
Cambridge University Press
Cambridge; New York; Melbourne 1985
352pp ISBN 0-521-24640-7
This is a full account of mineral, organic and
artificial substances, how their structures are
studied and how the knowledge of their properties
is put to ever greater use in our everyday material
world. Plenty of illustrations.

Back cover Pyrite specimen from Huansala,
Huanuco, Peru.